YOUR KNOWLEDGE HAS VALUE

- We will publish your bachelor's and master's thesis, essays and papers

- Your own eBook and book - sold worldwide in all relevant shops

- Earn money with each sale

Upload your text at www.GRIN.com and publish for free

Wind Tunnel. Theoretical and Mathematical Analysis

Abdusselam Šabić

Bibliographic information published by the German National Library:

The German National Library lists this publication in the National Bibliography; detailed bibliographic data are available on the Internet at http://dnb.dnb.de.

ISBN: 9783346707444
This book is also available as an ebook.

Contents

Introduction

I'll be exploring both the theoretical and experimental aspects of aerodynamic forces. I'll talk about the various atmospheric conditions and different forces affecting aircraft flight. I'll also include experimental data and calculations to explain my points.

Task 1

International Standard Atmosphere

Atmospheric properties

Atmospheric property	SI unit	English unit
Pressure	101.3 kPa	2116.7 lb_f/ft^2
Density	1.225 kgm^{-3}	0.002378 $slug/ft^3$
Temperature	288 K (15°C)	518 R (59°F)
Speed of sound	340.3 ms^{-1}	1116.4 ft/s
Viscosity	$1.789x10^{-5}$ $kgm^{-1}s^{-1}$	$3.737 x10^{-7}$
Kinematic viscosity	$1.460x10^{-5}$ m^2/s	$1.5723x10^{-4}$ ft^2/s
Thermal conductivity	0.02596 W/m/K	0.015 BTU/hr/ft/°R
Gas constant	287.1 J/Kg/K	1715.7 ft $lb_f/slug/°R$
Specific heat (C_p)	1005 J/Kg/K	6005 ft $lb_f/slug/°R$
Specific heat (C_v)	717.98 J/Kg/K	4289 ft $lb_f/slug/°R$
Ratio of specific heats	1.40	
Gravitational acceleration	9.80665 m/s^2	32.174 ft/s^2

(mdp.eng.cam.ac.uk, n.d.)

Standard atmosphere

The air in the atmosphere consists of 78% nitrogen, 21% oxygen, 0.93% argon and 0.04% carbon dioxide. The remaining 0.03% consists of "trace amounts of neon, helium, methane, krypton, hydrogen and water vapour". (Sharp, 2017) (space.com)

Layers of the atmosphere

There are 6 main layers of the atmosphere, and they are as follows

- Troposphere, which extends from the Earth's surface up to altitudes 8 to 14.5 km. Nearly all the Earth's weather is in this region.

- Stratosphere, which starts just above the troposphere and extends to an altitude of 50 km. The stratosphere contains 90% of the ozone layer.
- Mesosphere, which starts just above the stratosphere and extends up to an altitude of 85 km.
- Thermosphere, which starts above the mesosphere and extends to 600 km above the Earth's surface. Satellites orbit the Earth in this layer.
- Ionosphere, which is abundant in electrons and ionised atoms, and it starts from somewhere around 48 km above the Earth's surface to the edge of space at 965 km, thus overlapping into the thermosphere and mesosphere.
- Exosphere, which is the uppermost limit of the atmosphere, extending from the thermosphere up to 10 000 km. (nasa.gov, 2017)

Geo potential Altitude above Sea Level - h - (m)	Temperature - t - (°C)	Acceleration of Gravity - g - (m/s²)	Absolute Pressure - p - (10⁴ N/m²)	Density - ρ - (kg/m³)	Dynamic Viscosity - μ - (10⁻⁵ N s/m²)
-1000	21.50	9.810	11.39	1.347	1.821
0	15.00	9.807	10.13	1.225	1.789
1000	8.50	9.804	8.988	1.112	1.758
2000	2.00	9.801	7.950	1.007	1.726
3000	-4.49	9.797	7.012	0.9093	1.694
4000	-10.98	9.794	6.166	0.8194	1.661
5000	-17.47	9.791	5.405	0.7364	1.628
6000	-23.96	9.788	4.722	0.6601	1.595
7000	-30.45	9.785	4.111	0.5900	1.561
8000	-36.94	9.782	3.565	0.5258	1.527
9000	-43.42	9.779	3.080	0.4671	1.493
10000	-49.90	9.776	2.650	0.4135	1.458
15000	-56.50	9.761	1.211	0.1948	1.422
20000	-56.50	9.745	0.5529	0.08891	1.422
25000	-51.60	9.730	0.2549	0.04008	1.448
30000	-46.64	9.715	0.1197	0.01841	1.475
40000	-22.80	9.684	0.0287	0.003996	1.601
50000	-2.5	9.654	0.007978	0.001027	1.704
60000	-26.13	9.624	0.002196	0.0003097	1.584
70000	-53.57	9.594	0.00052	0.00008283	1.438
80000	-74.51	9.564	0.00011	0.00001846	1.321

Figure 1 US standard atmosphere (The Engineering ToolBox, n.d.)

Effects of changing altitude on the ISA values

In the troposphere (11 km altitude), temperature decreases linearly. The variation of temperature, pressure, density, viscosity and speed of sound can be calculated using the hydrostatic equation for a column of air, which is assumed to be in ideal conditions. The lapse rate of the atmosphere in the troposphere is 6.5 K/km.

The hydrostatic equation for the troposphere is

$$\frac{\delta p}{\delta h} = -\rho g$$

The ideal gas law is as follows

$$p = \rho RT$$

The lapse rate equation is as follows

$$T = T_0 - Lh$$

The variables used are as follows:

P – pressure (Pa)

T – temperature (K)

ρ – density (kg/m^3)

g – gravitational acceleration (9.81 m/s^2)

T_o – standard sea level temperature (288 K)

R – ideal gas constant (287 m^2/s^2/K)

h – altitude above sea level (m)

L – lapse rate (0.0065 K/m)

The pressure variation in the troposphere is obtained by the following equation:

$$\frac{p}{p_o} = \left(\frac{T}{T_0}\right)^{\frac{g}{LR}}$$

When the hydrostatic equation is solved with a constant temperature, the pressure variation in the stratosphere is obtained by the following equation:

$$\frac{p}{p_s} = e^{\left(\frac{-g(h_s-h)}{RT_s}\right)}$$

The subscript s on the values of p, h and T denotes the pressure, altitude and temperature at the end of the troposphere, i.e. the start of the stratosphere.

Where p_0 = 101.3 kPa. Density is calculated using the ideal gas law, and the dynamic viscosity is obtained by the Sutherland law, as follows:

$$\mu = \frac{1.458^{10^{-6}}\sqrt{T}}{1 + 110.4 \ / \ T}$$

Kinematic viscosity is calculated by dividing the dynamic viscosity by density:

$$v = \frac{\mu}{\rho}$$

Worked example:

Altitude of 3 km

T = 288 – 0.0065(3000) = 268.5 K

$$p = 101300 \left(\frac{268.5}{288}\right)^{\frac{9.81}{0.0065(287)}} = 70\ 064 \text{ Pa}$$

$$\rho = \frac{70064}{287(268.5)} = 0.909 \text{ kgm}^{-3}$$

$$\mu = \frac{1.458^{10^{-6}}\sqrt{268.5}}{1+110.4/268.5} = 1.693 \times 10^{-5} \text{ kgm}^{-1}\text{s}^{-1}$$

$$v = \frac{1.693^{10^{-5}}}{0.909} = 1.862 \times 10^{-5} \text{ m}^2\text{s}^{-1}$$

Altitude of 15 km

T = 288 – 0.0065(11000) = 216.5 K

$$p = p_s e^{\left(\frac{-g(h_s-h)}{RT_s}\right)} = 22588 e^{\left(\frac{-9.81(15000-11000)}{287(216.5)}\right)} = 12\ 012 \text{ Pa}$$

$$\rho_s = \frac{p_2}{RT} \Rightarrow \rho_s = \frac{12110}{287(216.5)} = 0.195 \text{ kgm}^{-3}$$

$$\mu = \frac{1.458^{10^{-6}}\sqrt{216.5}}{1+110.4/216.5} = 1.421 \times 10^{-5} \text{ kgm}^{-1}\text{s}^{-1}$$

$$v = \frac{1.421^{10^{-5}}}{0.195} = 7.29 \times 10^{-5}$$

Analysis of ISA properties with changing altitude

The range of values calculated here is for 2 different sections of the atmosphere, i.e. the troposphere and stratosphere respectively. The main reason for the variation between these values is that the air in the troposphere is heated and thus rises. As it rises it gets cooled down, and the higher you go the cooler the air becomes. As seen later on in the document, the lapse rate is 0.0065 K/m, or in other words 6.5 K/km, which means that for every km climbed, the air temperature decreases by 6.5 K.

This portion of the atmosphere is a linear graph with a negative gradient, meaning that a set of equations can be determined for this region. In addition, density is directly proportional to pressure and inversely proportional to temperature, which can be seen in the pressure equation for the troposphere by rearranging the values.

The reason that temperature is stable in the lower part of the stratosphere (from 11 km to 20 km altitude) is due to cooler, denser air being trapped underneath the warmer, less dense air. This results in very low turbulence in this region.

Density decreases with height because there's less air pushing down from above, and gravity is weaker further away from the Earth's surface. Since density is equal to the air's mass divided by its volume, the lower gravity results in decreased mass and increased volume.

The reason for changing pressure is that as the altitude increases, the weight of the air decreases due to the fact that there are less air molecules above a certain surface compared to a similar surface at lower altitudes. In other words, since pressure is dependent on both density and temperature, and because both of them decrease, the pressure automatically decreases as well.

Dynamic viscosity is dependent only on temperature, so just like temperature, it decreases steadily until 11 km altitude and stays constant between 11 and 20 km altitude.

The reason that there are 2 sets of equations is that up to 11 km, temperature decreases, hence lapse rate is a factor, whereas from 11 to 20 km lapse rate is zero, so new equations must be formed.

Bernoulli's equation and Euler's equation

In general terms, Bernoulli's equation is as follows: static pressure + dynamic pressure = total pressure. This equation is constant as follows:

$$p_1 + \rho g h_1 + \frac{1}{2}\rho v_1^2 = p_2 + \rho g h_2 + \frac{1}{2}\rho v_2^2$$

In a wind tunnel experiment, density is constant, h is constant, and g is constant as well, so these values cancel out.

$$p_1 + \frac{1}{2}\rho v_1^2 = p_2 + \frac{1}{2}\rho v_2^2$$

Euler's equation for steady flow along a streamline is the relation between the velocity, pressure and density of a moving fluid. Integrating the equation gives rise to Bernoulli's equation in the form of energy per unit weight of the following fluid. These equations depend on the (x,y) coordinates, velocity components (u,v), pressure and density.

2D steady form equations:

Continuity $\dfrac{\delta(\rho u)}{\delta x} + \dfrac{\delta(\rho v)}{\delta y} = 0$

X-momentum $\dfrac{\delta(\rho u^2)}{\delta x} + \dfrac{\delta(\rho uv)}{\delta y} = -\dfrac{\delta p}{\delta x}$

Y-momentum $\dfrac{\delta(\rho uv)}{\delta x} + \dfrac{\delta(\rho v^2)}{\delta y} = -\dfrac{\delta p}{\delta y}$

Incompressible form

Continuity $\dfrac{\delta u}{\delta x} + \dfrac{\delta v}{\delta y} = 0$

X-momentum $u\dfrac{\delta u}{\delta x} + v\dfrac{\delta u}{\delta y} = -\dfrac{1}{\rho}\dfrac{\delta p}{\delta x}$

Y-momentum $u\dfrac{\delta v}{\delta x} + \dfrac{\delta v}{\delta y} = -\dfrac{1}{\rho}\dfrac{\delta p}{\delta y}$

Lift and drag affecting flight

Lift and drag are forces that have both magnitude and direction, i.e. they're vector forces. Lift acts upwards perpendicularly to the wing surface in level flight and drag is a mechanical force generated by a body propagating through a fluid, and it acts in the opposite direction to thrust. In level flight, lift = weight and thrust = drag. Drag is generated by every part and surface of an aircraft, whereas lift is only generated by the wings.

The higher the aircraft velocity, the more lift is generated. There are 3 main types of drag: parasitic, induced and wave drag. Parasitic drag consists of form drag (aerodynamic resistance to motion due to the aircraft shape), skin friction (caused by smoothness/roughness of aircraft surfaces) and interference drag (when varying characteristics meet, ex. wind and fuselage). Induced drag is "a secondary effect of the production of lift" and wave drag occurs during shock wave formation.

"Parasitic drag increases with the square of the airspeed, while induced drag, being a function of lift, is greatest when maximum lift is being developed, usually at low speeds." (SkyBrary, 2017) Parasitic drag increases with altitude and induced drag increases with altitude.

Lift on the other hand arises as a result of differences in air pressure. The lower aerofoil surface has less area and thus higher pressure because airflow is faster here, and the upper surface is bent more so has a larger surface area and thus lower pressure. Additionally, as the angle of attack increases, lift generated also increases as a result of bigger pressure gradient until the stall angle is reached, where the coefficient of lift exactly equals the coefficient of drag. After that, C_D increases and stalling occurs, which results in loss of lift and falling of the aircraft if not corrected.

How to obtain lift and drag forces

Lift is obtained by the following equation $L = \frac{1}{2}c_L \rho S v^2$

Drag is obtained by the following equation $D = \frac{1}{2}c_D \rho S v^2$

Wind tunnel experiment

The wind tunnel experiment is done for the purpose of analysing aerodynamic forces, namely lift and drag, by comparing the values obtained from the experiment with standard values obtained by aerodynamicists worldwide.

There are 3 sets of experiments for analysing the aerodynamic forces, and they compare 2 aerofoil sections with different flap angles, 2 different velocities and 2 different angles of attack. These experiments look at how C_L and C_D change when these variables are varied at times and kept constant at other times.

Procedure

Item	Details
Dimensions	300 mm span and 152 mm chord
Profile	NACA 0012
Weight	600 g

<u>**Wind tunnel operation**</u>

a. Start up
- Switch on the electrical isolator on the control and instrumentation frame.
- Set the speed control to the minimum position (fully anticlockwise)
- Press the green "START" button
- **<u>Gradually</u>** turn the speed control clockwise until the tunnel is operation at the speed needed for the experiment

b. Shut down
- Slowly turn the speed control fully anticlockwise
- Press the red STOP button.

c. Emergency Stop
- Press the red STOP button
- Turn the speed control fully anticlockwise

Model installation

1. Make sure the electrical supply to the wind tunnel is disconnected.

2. Remove the side window opposite the three component balance.

3. Tighten the balance locks of the three component balance. From outside the wind tunnel, insert one of the airfoils into the collect of the three component balance, so that its support shaft passes into the wind tunnel test section

4. Inside the test section, measure the distance from the centre of the airfoil to the bottom surface of the test section. This is nominally 152.5mm. Remove the airfoil

5. From inside the test section, insert the airfoil into the model clamp of the balance, so that its leading edge faces into the airflow. Set the angle of the three component balance scale to zero degrees. Do not tighten the model clamp yet.

6. Adjust the trailing edge of the aerofoil to the same height 152.5mm and tighten the model clamp of the balance. The airfoil is now set as zero angle of attack.

7. Use the angle adjustment of the balance to rotate the airfoil and check that it rotates freely in the test section, without rubbing the sides of the test section. Experimental result will be spoiled if the edges (wingtips) of the airfoil touch the sides of the test section during experiment.

8. Recheck the adjustments.

9. Refit the side window of the test section

<u>**Wind Tunnel test procedure**</u>

1. Record the ambient temperature and pressure.

2. <u>**Release balance lock before test**</u>

3. Start the wind tunnel, set free-stream velocity as assigned to each team.

4. Record the Lift, Drag and Pitching moment readings from the Three component balance display unit

5. Adjust the angle of attack in steps of $1°$ from $-5°$ upwards until the lift passed its maximum magnitude due to the stall.

6. The airfoil is flown "upside down" in the three component balance. This is normal and allows more accurate balance design. (The sign of angle of attack in this balance is the opposite which appears in your text book)

7. Repeat procedure with other NACA _ _ aerofoil

8. <u>**Tighten balance lock after test.**</u>

Experiment 1

In this experiment, the windspeed is kept constant (dependent variable) and the angle of attack is changed (independent variable). The experiment is repeated 2 times, at 20 m/s and 15 m/s windspeed respectively.

Data Series 1															
AFA2 Basic Balan	AFA3 Balance			Manual Angle Input	Operating Conditions					Aerofoil Characteristics					
Force	Lift	Drag	Pitching Moment	Manual Angle	Atmospheric Temperature	Atmospheric Pressure	Ambient Air Density	Calculated Windspeed	Aerofoil Span	Aerofoil Chord Length	Coefficient of Lift, C_L	Coefficient of Drag, C_D	Pitching Moment Coefficient	Avg of Cm	Lift / Drag Ratio, C_L / C_D
(N)	(N)	(N)	(Nm)	(Degrees)	(°C)	(mbar)	(kg.m^{-3})	(m.s^{-1})	(mm)	(mm)					
--	0.05	0.01	0	0	15	1013	1.23	19.62	300	152	0	0	0		--
--	-0.01	-0.01	0	0	15	1013	1.23	19.71	300	152	0	0	0	0	--
--	-0.02	-0.01	0	0	15	1013	1.23	19.71	300	152	0	0	0		--
--	-5.7	8.3	0.03	3	15	1013	1.23	20.12	300	152	0	0.73	0.02		-0.68
--	-5.7	8.29	0.03	3	15	1013	1.23	20.12	300	152	0	0.73	0.02	0.02	-0.68
--	-5.71	8.29	0.03	3	15	1013	1.23	20.08	300	152	0	0.73	0.02		-0.68
--	-2.41	7.55	0.03	6	15	1013	1.23	19.83	300	152	0	0.68	0.02		-0.32
--	-2.37	7.55	0.03	6	15	1013	1.23	19.87	300	152	0	0.68	0.02	0.02	-0.31
--	-2.38	7.55	0.03	6	15	1013	1.23	19.83	300	152	0	0.68	0.02		-0.32
--	0.58	7.35	0.02	9	15	1013	1.23	19.71	300	152	0.05	0.67	0.01		0.07
--	0.57	7.33	0.02	9	15	1013	1.23	19.67	300	152	0.05	0.68	0.01	0.01	0.07
--	0.53	7.32	0.02	9	15	1013	1.23	19.79	300	152	0.05	0.67	0.01		0.07
--	2.43	7.37	0.04	12	15	1013	1.23	19.37	300	152	0.23	0.7	0.03		0.33
--	2.43	7.37	0.05	12	15	1013	1.23	19.42	300	152	0.23	0.7	0.03	0.02666667	0.33
--	2.39	7.38	0.04	12	15	1013	1.23	19.42	300	152	0.23	0.7	0.02		0.33
--	5.53	7.37	0.03	15	15	1013	1.23	19.2	300	152	0.53	0.71	0.02		0.75
--	5.44	7.37	0.03	15	15	1013	1.23	19.2	300	152	0.53	0.71	0.02	0.02	0.75
--	5.35	7.37	0.03	15	15	1013	1.23	19.2	300	152	0.52	0.71	0.02		0.73
--	7.2	7.21	0	18	15	1013	1.23	19.25	300	152	0.69	0.69	0		1
--	7.19	7.2	0	18	15	1013	1.23	19.25	300	152	0.69	0.69	0	0	1
--	7.14	7.2	0	18	15	1013	1.23	19.25	300	152	0.69	0.69	0		1
--	8.83	7.21	0.02	21	15	1013	1.23	18.95	300	152	0.88	0.72	0.01		1.22
--	8.91	7.21	0.02	21	15	1013	1.23	18.99	300	152	0.88	0.71	0.01	0.01	1.24
--	8.96	7.19	0.02	21	15	1013	1.23	18.99	300	152	0.89	0.71	0.01		1.25
--	9.96	7.44	0.01	24	15	1013	1.23	18.99	300	152	0.96	0.74	0.01		1.32
--	9.98	7.45	0.01	24	15	1013	1.23	18.9	300	152	1	0.74	0.01	0.01	1.35
--	10.05	7.44	0.01	24	15	1013	1.23	19.08	300	152	0.98	0.73	0.01		1.34
--	10.43	7.96	-0.03	27	15	1013	1.23	18.9	300	152	1.04	0.79	0		1.32
--	10.27	7.88	-0.02	27	15	1013	1.23	18.99	300	152	1.02	0.78	0	0	1.31
--	10.34	7.76	-0.01	27	15	1013	1.23	19.08	300	152	1.01	0.76	0		1.33
--	10.74	7.57	0.01	30	15	1013	1.23	18.77	300	152	1.09	0.77	0.01		1.42
--	10.72	7.58	0.01	30	15	1013	1.23	18.86	300	152	1.07	0.76	0.01	0.01	1.41
--	10.72	7.61	0.01	30	15	1013	1.23	18.86	300	152	1.07	0.76	0.01		1.41
--	9.13	7.5	0	33	15	1013	1.23	19.25	300	152	0.88	0.72	0		1.22
--	9.27	7.51	0	33	15	1013	1.23	19.08	300	152	0.91	0.74	0	0	1.23
--	9.45	7.51	0	33	15	1013	1.23	19.12	300	152	0.92	0.73	0		1.26

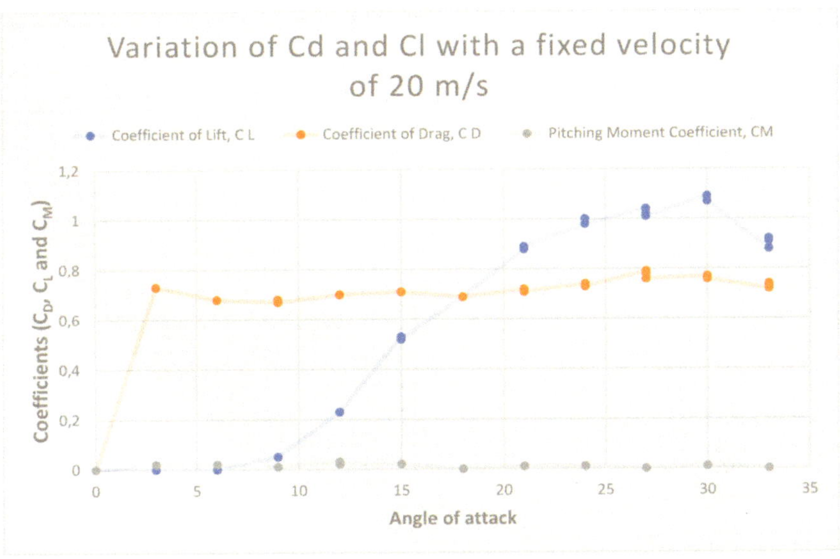

In this graph we can see that for angles from 0 to 6 degrees, C_L stays at zero, and C_D increases linearly. After that, the C_D starts to fluctuate a bit, but is more or less horizontal, i.e. it roughly levels off and doesn't change much when the angle of attack is increased.

C_L starts increasing slightly after 6 degrees, and C_D decreases slightly and stays between 0.65 and 0.8 throughout the rest of the experiment. C_L continues increasing linearly until an angle of attack of 15 degrees, and once again until 21 degrees.

At 18 degrees, C_L and C_D are exactly the same, i.e. 18 is the stall angle when the velocity is 20 m/s. In other words, zero lift is generated because the lift and drag are exactly equal to each other, thus cancelling each other out.

From 18 degrees onward, C_L keeps on increasing until an angle of attack of 30 degrees, after which it decreases. At this angle, C_L is 1.4, which shows that 30 degrees is the optimum angle of attack at a velocity of 20 m/s.

We can also see that the pitching moment coefficient is very small throughout the experiment, which means that stability around the pitch axis is high, or in other words the aerofoil doesn't pitch virtually at all.

	AFA2 Basic Balance	AFA3 Balance			AFA4 Encoder Input	Manual Angle Input	Operating Conditions					Aerofoil Characteristics					
Time																	
Time	Force	Lift	Drag	Pitching Moment	Angle	Manual Angle	Atmospheric Temperature	Atmospheric Pressure	Ambient Air Density	Calculated Windspeed	Aerofoil Span	Aerofoil Chord Length	Coefficient of Lift, C_L	Coefficient of Drag, C_D	Pitching Moment Coefficient, C_M	Lift / Drag Ratio, C_L / C_D	
(s)	(N)	(N)	(N)	(Nm)	(Degrees)	(Degrees)	(°C)	(mbar)	(kg.m⁻³)	(m.s⁻¹)	(mm)	(mm)					
--	--	-0.01	-0.06	0	0.2	0	15	1013	1.23	15.06	300	152	0	0	0	0	
--	--	0	-0.07	0	0.2	0	15	1013	1.23	15.06	300	152	0	0	0	0	
--	--	0.01	-0.08	0	0.2	0	15	1013	1.23	15.12	300	152	0	0	0	0	
--	--	0.09	-0.01	0.03	3.4	3	15	1013	1.23	15.01	300	152	0.01	0	0.03	--	
--	--	0.09	-0.03	0.03	3.4	3	15	1013	1.23	15.01	300	152	0.01	0	0.03	--	
--	--	0.09	-0.03	0.03	3.4	3	15	1013	1.23	14.95	300	152	0.01	0	0.03	--	
--	--	3.82	0.77	-0.14	6.4	6	15	1013	1.23	14.9	300	152	0.61	0.12	-0.15	5.08	
--	--	3.76	0.72	-0.14	6.4	6	15	1013	1.23	14.9	300	152	0.6	0.12	-0.15	5	
--	--	3.75	0.76	-0.14	6.4	6	15	1013	1.23	14.95	300	152	0.6	0.12	-0.15	5	
--	--	4.68	0.79	-0.23	9.7	9	15	1013	1.23	14.79	300	152	0.76	0.13	-0.25	5.85	
--	--	4.71	0.8	-0.23	9.7	9	15	1013	1.23	14.79	300	152	0.77	0.13	-0.25	5.92	
--	--	4.72	0.8	-0.23	9.7	9	15	1013	1.23	14.73	300	152	0.78	0.13	-0.25	6	
--	--	8.36	1.85	-0.42	12.6	12	15	1013	1.23	14.68	300	152	1.38	0.31	-0.46	4.45	
--	--	8.27	1.85	-0.42	12.6	12	15	1013	1.23	14.68	300	152	1.37	0.31	-0.46	4.42	
--	--	8.24	1.85	-0.42	12.6	12	15	1013	1.23	14.62	300	152	1.37	0.31	-0.46	4.42	
--	--	4.24	2.21	-0.17	15.6	15	15	1013	1.23	14.9	300	152	0.68	0.35	-0.18	1.94	
--	--	4.28	2.22	-0.17	15.6	15	15	1013	1.23	14.73	300	152	0.7	0.36	-0.18	1.94	
--	--	4.24	2.22	-0.17	15.6	15	15	1013	1.23	14.68	300	152	0.7	0.37	-0.19	1.89	

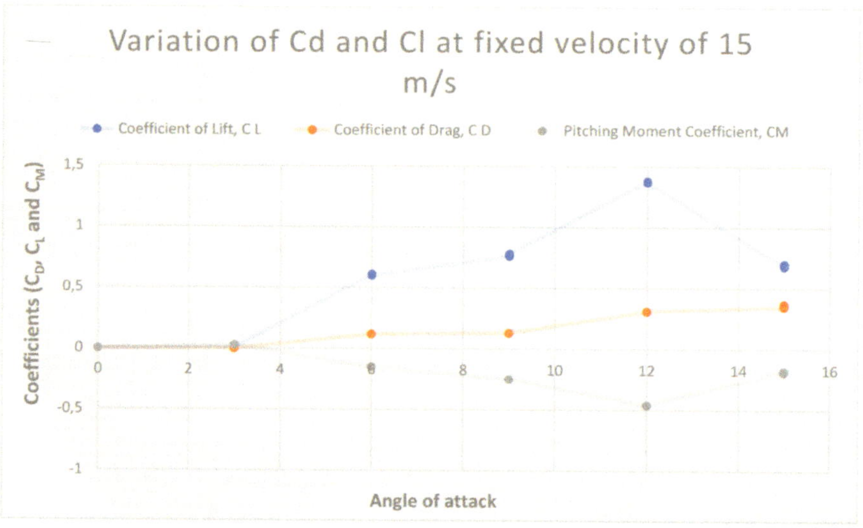

In this graph, it's evident that both C_L and C_D are equal to zero from 0 to 3 degrees angle of attack. However, unlike at 20 m/s, the C_L value is always larger than the C_D value at 15 m/s. What this means is that from 0 to 15 degrees angle of attack, there's always positive lift and minimal drag.

15

We can also see that from 3 to 6 degrees angle of attack, both values increase linearly, but the gradient of C_L is much larger, therefore it increases at a bigger rate. After that, from 6 to 9 degrees, C_L increases linearly as well, but this time at a lower gradient. At this stage, C_D levels off and stays at the same value.

From 9 to 12 degrees, C_L increases in a similar way that it did from 3 to 6 degrees. C_D also increases but at a smaller rate. It's evident from this graph that at a velocity of 15 m/s, the greatest lift generated is at an angle of attack of 12 degrees, because the difference between the 2 values is the largest here.

After 12 degrees, C_L decreases rapidly, and C_D increases a bit more. It's evident that at this stage less lift is being produced and more drag is being produced. It should also be noted that this experiment was stopped earlier than the previous one; this is because it's already evident that C_L will decrease rapidly, and since the optimum angle of attack is already known, there was no need to continue the experiment further.

In this graph, C_M is zero at angles of attack from 0 to 3 degrees, which signifies level flight (movement about the pitch axis is zero). After this, C_M starts decreasing until an angle of attack of 12 degrees, and increases slightly until 15 degrees, while remaining under zero at all times, which simply means that the aerofoil section being tested is upside down. There's no conflict in this, as the aerofoil section was indeed placed upside down during the experiment. This was done so that the value of lift obtained will be positive, since the wind tunnel measuring instruments measure negative lift, i.e. placing the aerofoil upside down cancels out the negative sign in the lift and returns a positive value.

Comparison between the two graphs

0 to 3 degrees:

At 20 m/s, C_D increases rapidly in a linear fashion, and C_L is zero.

At 15 m/s, the thrust produced is not enough to generate lift at such low angles of attack, so both values are approximately zero.

3 to 6 degrees:

At 20 m/s, C_D decreases slightly, and C_L is still zero.

At 15 m/s, both of them increase but C_L increases more rapidly.

Both happen in a linear fashion at this stage.

6 to 12 degrees:

At 20 m/s, C_D stays approximately the same and C_L increases by a large amount.

At 15 m/s, there are 2 distinct stages, as mentioned earlier. From 6 to 9 degrees, C_L increases and C_D stays the same, and in the second stage both increase, but C_D increases less than C_L.

The rest of the differences were explained earlier. It should be noted that in the first graph, the stall angle is known because more values of angle of attack were used, whereas in the second experiment the angle of attack was only increased up to 15 degrees.

However, the second graph can be extrapolated to estimate when the stall angle is reached. While it's not accurate because we don't know whether the gradient changes or not, it can still provide us with reasonable values and low error percentage if done correctly. For example, provided that the gradients of the lines in the final part of the graph stay constant, then the stall angle can be estimated to be somewhere around 16 degrees.

Experiment 2

In this experiment, The NACA2412 aerofoil was experimented on in the wind tunnel at 0 degrees flap and 22 degrees flap and a fixed velocity of 10 m/s in both cases, as well as varying angles of attack.

Data Series 1												
AFA3 Balance			Manual Angle Input	Operating Conditions					Aerofoil Characteristics			
Lift	Drag	Pitching Moment	Manual Angle	Atmospheric Temperature	Atmospheric Pressure	Ambient Air Density	Calculated Windspeed	Coefficient of Lift, C_L	Coefficient of Drag, C_D	Pitching Moment Coeffici	Lift / Drag Ratio, C_L / C_D
(N)	(N)	(Nm)	(Degrees)	(°C)	(mbar)	(kg.m^{-3})	(m.s^{-1})				
0.13	0.82	0	0	15	1013	1.23	10.06	0.05	0.29	0	0.17
0.09	0.82	0	0	15	1013	1.23	9.9	0.03	0.3	0	0.1
0.07	0.83	0	0	15	1013	1.23	9.9	0.03	0.3	0	0.1
0.17	3.43	0.01	3	15	1013	1.23	10.06	0.06	1.21	0.02	0.05
0.2	3.43	0.01	3	15	1013	1.23	10.14	0.07	1.19	0.02	0.06
0.22	3.44	0.01	3	15	1013	1.23	10.06	0.08	1.21	0.02	0.07
0.6	3.9	0.01	6	15	1013	1.23	10.06	0.21	1.37	0.02	0.15
0.61	3.9	0.01	6	15	1013	1.23	10.06	0.21	1.37	0.02	0.15
0.59	3.91	0.01	6	15	1013	1.23	10.14	0.2	1.36	0.02	0.15
1.75	3.93	-0.08	9	15	1013	1.23	10.06	0.62	1.38	-0.19	0.45
1.77	3.92	-0.08	9	15	1013	1.23	10.06	0.62	1.38	-0.19	0.45
1.75	3.92	-0.08	9	15	1013	1.23	10.06	0.62	1.38	-0.19	0.45
1.88	3.65	-0.1	12	15	1013	1.23	9.9	0.68	1.33	-0.24	0.51
1.91	3.64	-0.1	12	15	1013	1.23	10.06	0.67	1.28	-0.23	0.52
1.94	3.63	-0.1	12	15	1013	1.23	9.98	0.69	1.3	-0.24	0.53
1.71	3.39	-0.09	15	15	1013	1.23	10.14	0.59	1.18	-0.21	0.5
1.71	3.39	-0.09	15	15	1013	1.23	10.14	0.59	1.18	-0.21	0.5
1.72	3.4	-0.08	15	15	1013	1.23	9.98	0.62	1.22	-0.19	0.51
1.62	3	-0.07	18	15	1013	1.23	9.98	0.58	1.07	-0.16	0.54
1.62	2.99	-0.07	18	15	1013	1.23	9.9	0.59	1.09	-0.17	0.54
1.66	2.98	-0.08	18	15	1013	1.23	10.06	0.58	1.05	-0.19	0.55

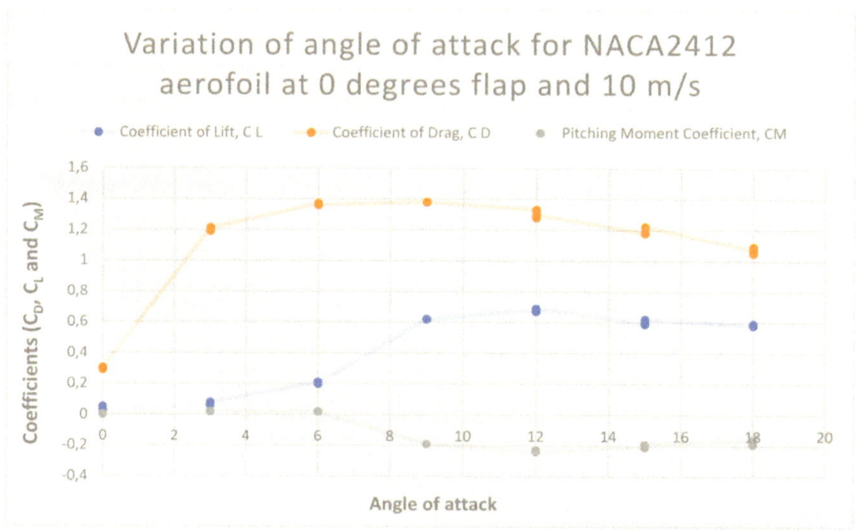

First off, we can see that at an angle of attack of 0, C_D is 0.3 and C_L is just slightly above zero. Then, from 0 to 3 degrees angle of attack, C_D increases rapidly in a linear fashion to a value of 1.2, and C_L barely increases at all and stays below 0.1.

From 3 to 6 degrees, both values increase linearly at a very similar rate, i.e. their gradients in this section of the graph are almost, if not, identical.

After this, we can see that C_D slowly starts levelling off and then starts decreasing slightly until 12 degrees, then more after that, whereas C_L increases rapidly from 6 to 9 degrees and slightly from 9 to 12 degrees, both linearly. To an angle of 15 degrees, C_L decreases to the same value as it was at 9 degrees and it levels off at this stage, staying constant.

C_D is always larger than C_L, i.e. there's always more drag produced than there is lift. This shows that a flap angle of zero doesn't provide sufficient surface area for generating lift.

C_M is zero until an angle of attack of 6 degrees, which means that the pitching attitude was zero. After that, C_M decreases to a negative value, which implies that the angle of attack of the aerofoil was increasing in a negative direction (downwards).

Data Series 1											
Time	AFA3 Balance			Manual Angle Input	Operating Conditions				Aerofoil Characteristics		
Time	Lift	Drag	Pitching Moment	Manual Angle	Atmospheric Temperature	Atmospheric Pressure	Ambient Air Density	Calculated Windspeed	Coefficient of Lift, C_L	Coefficient of Drag, C_D	Pitching Moment Coefficient, C_M	Lift / Drag Ratio, C_L / C_D
(s)	(N)	(N)	(Nm)	(Degrees)	(°C)	(mbar)	(kg.m^{-3})	(m.s^{-1})				
--	-0.03	0.06	0	0	15	1013	1.23	9.98	0	0.02	0	-0.5
--	-0.04	0.07	0	0	15	1013	1.23	9.98	0	0.03	0	-0.33
--	-0.03	0.07	0	0	15	1013	1.23	9.9	0	0.03	0	-0.33
--	1.29	1	-0.04	3	15	1013	1.23	9.81	0.48	0.37	-0.1	1.3
--	1.29	1	-0.04	3	15	1013	1.23	9.81	0.48	0.37	-0.1	1.3
--	1.27	1.01	-0.04	3	15	1013	1.23	9.73	0.48	0.38	-0.1	1.26
--	5.53	0.88	-0.34	6	15	1013	1.23	9.64	2.12	0.34	-0.86	6.24
--	5.51	0.88	-0.34	6	15	1013	1.23	9.64	2.11	0.34	-0.86	6.21
--	5.48	0.88	-0.34	6	15	1013	1.23	9.64	2.1	0.34	-0.86	6.18
--	4.28	1.48	-0.3	9	15	1013	1.23	9.39	1.73	0.6	-0.8	2.88
--	4.33	1.47	-0.3	9	15	1013	1.23	9.39	1.75	0.59	-0.8	2.97
--	4.5	1.47	-0.28	9	15	1013	1.23	9.47	1.79	0.58	-0.73	3.09

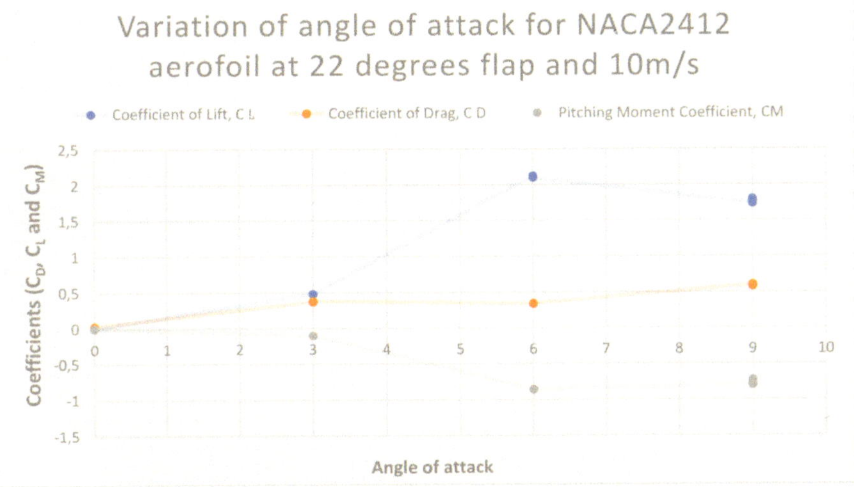

We can see that both values increase linearly from 0 to 3 degrees angle of attack, and C_L is only slightly larger than C_D. This means that unlike the first case, increasing the flap angle results in having to increase the angle of attack as well.

This is seen in the graph when the lines are very close to each other, i.e. lift is almost the same as drag. It's only after the angle of attack is increased to 6 that we see a much better improvement than in the previous scenario.

This happens because C_L increases to a much larger value of 2.1 at 6 degrees, and C_D decreases very slightly. While it's true that C_D is larger at this angle than it was in the previous case, the much larger C_L value mitigates for this by providing a larger difference between the 2 values, which means that more lift is generated.

From 6 to 9 degrees, C_L decreases slightly to 1.75, whereas C_D increases slightly by roughly 0.15 to 0.2. Based on these results, it can be concluded that 6 degrees is the optimum angle of attack for a flap angle of 22 degrees.

In this graph, we can see that changing the flap angle increases the C_L value for increasing angle of attack. While C_L increases by a larger amount than for a flap angle of 0, C_D doesn't change much, with only a 0.1 increase at an angle of attack of 9 degrees compared to the first graph in this experiment.

Here C_M is also negative, which implies that the pitch attitude was negative, or nose-down.

Experiment 3

In this experiment, the angle of attack is fixed and the windspeed is the independent variable. The experiment is repeated 2 times at 2 different values of angle of attack, namely 10 degrees and 30 degrees respectively, while everything else is kept constant.

	AFA2 Basic Balance	AFA3 Balance			AFA4 Encoder Input	Manual Angle Input	Operating Conditions					Aerofoil Characteristics								
Time	Force	Lift	Drag	Pitching Moment	Angle	Manual Angle	Atmospheric Temperature	Atmospheric Pressure	Ambient Air Density	Calculated Windspeed	Average of windspeed	Aerofoil Span	Aerofoil Chord Length	Coefficient of Lift, C_L	Average of CL	Coefficient of Drag, C_D	Average of CD	Pitching Moment Coefficient	Average of CM	Lift/Drag Ratio, C_L/C_D
(s)	(N)	(N)	(N)	(Nm)	(Degrees)	(Degrees)	(°C)	(mbar)	(kg.m^{-3})	(m.s^{-1})	(m/s)	(mm)	(mm)							
--	--	0.08	0.04	0	10.4	10	15	1013	1.23	0		300	152	--		--		--		--
--	--	0.08	0.05	0	10.4	10	15	1013	1.23	0	0.428066667	300	152	--	1.74	--	1.09	--	0	--
--	--	0.08	0.05	0	10.4	10	15	1013	1.23	1.28		300	152	1.74		1.09		0		1.6
--	--	0.08	0.06	0	10.4	10	15	1013	1.23	3.36		300	152	0.25		0.25		0		1
--	--	0.08	0.09	0	10.4	10	15	1013	1.23	3.38	3.38	300	152	0.25	0.23	0.28	0.27	0	0	0.89
--	--	0.06	0.09	0	10.4	10	15	1013	1.23	3.38		300	152	0.19		0.28		0		0.68
--	--	0.09	0.04	0	10.4	10	15	1013	1.23	5.71		300	152	0.1		0.04		0		2.5
--	--	0.14	0.06	0	10.4	10	15	1013	1.23	5.85	5.803333333	300	152	0.15	0.12	0.06	0.046666667	0	0	2.5
--	--	0.11	0.04	0	10.4	10	15	1013	1.23	5.85		300	152	0.11		0.04		0		2.75
--	--	0.25	0.04	0	10.4	10	15	1013	1.23	8.94		300	152	0.11		0.02		0		5.5
--	--	0.29	0.05	0	10.4	10	15	1013	1.23	8.94	8.94	300	152	0.13	0.12	0.02	0.02	0	0	6.5
--	--	0.26	0.05	0	10.4	10	15	1013	1.23	8.94		300	152	0.12		0.02		0		6
--	--	0.27	-0.1	0	10.4	10	15	1013	1.23	11.98		300	152	0.07		-0.02		0		-3.5
--	--	0.24	-0.1	0	10.4	10	15	1013	1.23	11.98	11.93666667	300	152	0.08	0.07	-0.02	-0.023333333	0	0	-3
--	--	0.3	-0.11	0	10.4	10	15	1013	1.23	11.85		300	152	0.08		-0.03		0		-2.67
--	--	0.31	-0.38	0.01	10.4	10	15	1013	1.23	15.01		300	152	0.05		-0.06		0.01		-0.83
--	--	0.31	-0.38	0.01	10.4	10	15	1013	1.23	15.08	15.04333333	300	152	0.05	0.05	-0.06	-0.06	0.01	0.01	-0.83
--	--	0.3	-0.39	0.01	10.4	10	15	1013	1.23	15.06		300	152	0.05		-0.06		0.01		-0.83
--	--	0.36	-1.06	0.01	10.4	10	15	1013	1.23	17.98		300	152	0.04		-0.12		0.01		-0.33
--	--	0.32	-1.05	0.01	10.4	10	15	1013	1.23	17.79	17.82333333	300	152	0.04	0.039666667	-0.12	-0.12	0.01	0.01	-0.33
--	--	0.29	-1.08	0.01	10.4	10	15	1013	1.23	17.7		300	152	0.03		-0.12		0.01		-0.25
--	--	0.51	-1.82	0.01	10.4	10	15	1013	1.23	20.83		300	152	0.04		-0.15		0.01		-0.27
--	--	0.54	-1.82	0.01	10.4	10	15	1013	1.23	20.83	20.88333333	300	152	0.04	0.043333333	-0.15	-0.15	0.01	0.01	-0.27
--	--	0.61	-1.82	0.01	10.4	10	15	1013	1.23	20.99		300	152	0.05		-0.15		0.01		-0.33
--	--	0.51	-3	0	10.4	10	15	1013	1.23	24.1		300	152	0.03		-0.18		0		-0.17
--	--	0.53	-2.99	0	10.4	10	15	1013	1.23	23.97	23.97866667	300	152	0.03	0.03	-0.19	-0.186666667	0	0	-0.16
--	--	0.48	-2.98	0	10.4	10	15	1013	1.23	23.86		300	152	0.03		-0.19		0		-0.16
--	--	0.42	-3.24	0	10.4	10	15	1013	1.23	26.95		300	152	0.02		-0.16		0		-0.12
--	--	0.4	-3.23	0	10.4	10	15	1013	1.23	26.95	26.99	300	152	0.02	0.02	-0.16	-0.16	0	0	-0.12
--	--	0.35	-3.22	0	10.4	10	15	1013	1.23	27.07		300	152	0.02		-0.16		0		-0.12

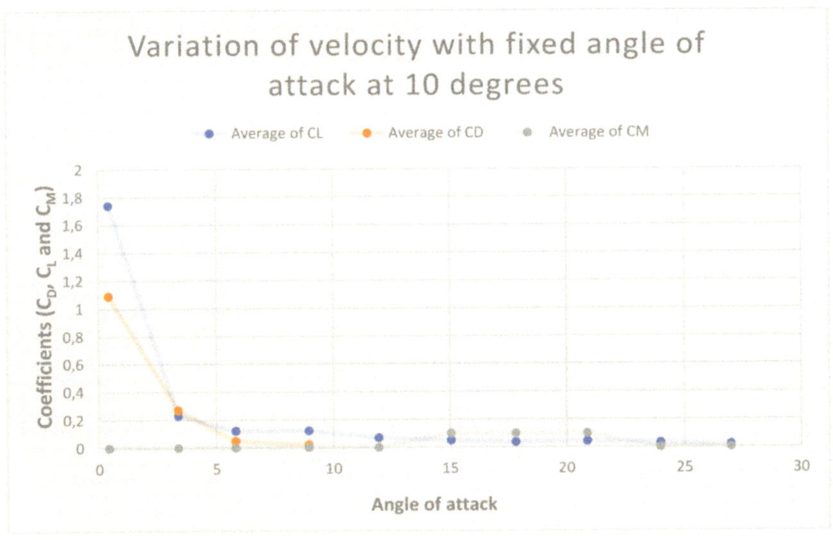

Based on the graph for an angle of attack of 10 degrees, it's evident that when windspeed is very low, the values of both C_L and C_D are relatively large, where C_L is larger than C_D. However, between windspeeds 0 and 3 m/s, both values decrease rapidly in a linear way to a very low value of 0.2. While this is preferable for C_D, it's not preferred for C_L.

Because both values are the same at a windspeed of 3 m/s, there is no lift generated and the aerofoil stalls. At windspeed of 6 m/s, both values decrease to nearly zero and level off until a windspeed of 9 m/s. At this stage, C_D stays at zero for the rest of the experiment, and C_L continues to decrease slightly, as shown by the blue part of the graph being almost horizontal.

Based on this experiment, it's evident that an angle of attack of 10 is too low to generate any lift at higher windspeeds. However, in this experiment, C_M is positive, which means that the pitching attitude was also positive, i.e. the pitch stability was high during this experiment.

AFA2 Basic Balan	AFA3 Balance			Manual Angle	Operating Conditions					Aerofoil Characteristics								
Force	Lift	Drag	Pitching Moment	Manual Angle	Atmospheric Temperature	Atmospheric Pressure	Ambient Air Density	Calculated Windspeed	Aerofoil Span	Aerofoil Chord Length	Coefficient of Lift, C_L	Average of CL	Coefficient of Drag, C_D	Average of CD	Pitching Moment Coeffi	Average of CM	Lift / Drag Ratio, C_L / C_D
(N)	(N)	(N)	(Nm)	(Degrees)	(°C)	(mbar)	(kg.m⁻¹)	(m.s⁻¹)	(mm)	(mm)							
--	0	-0.01	0	33	15	1013	1.23	0	300	152	0		0		--		--
--	0	-0.01	0	33	15	1013	1.23	0	300	152	0	0	0		--	0	--
--	0	-0.01	0	33	15	1013	1.23	0	300	152	0		0	0	--		--
--	0.04	-0.01	0	33	15	1013	1.23	1.81	300	152	0.44		0		0		-4
--	0.04	-0.01	0	33	15	1013	1.23	1.81	300	152	0.44	0.39	0		0	0	-4
--	0.04	-0.01	0	33	15	1013	1.23	2.21	300	152	0.29		0	0	0		-4.14
--	0.17	0.01	0	30	15	1013	1.23	3.83	300	152	0.41		0.02		0		20.5
--	0.18	0.01	0	30	15	1013	1.23	3.83	300	152	0.44	0.413333333	0.02		0	0	22
--	0.18	0.01	0	30	15	1013	1.23	4.04	300	152	0.39		0.02	0.02	0		19.5
--	0.45	0.03	0	30	15	1013	1.23	5.85	300	152	0.47		0.03		0		15.67
--	0.45	0.03	0	30	15	1013	1.23	5.85	300	152	0.47	0.47	0.03		0	0	15.67
--	0.45	0.03	0	30	15	1013	1.23	5.85	300	152	0.47		0.03	0.03	0		15.67
--	0.81	0.07	0	30	15	1013	1.23	7.98	300	152	0.45		0.04		0		11.25
--	0.79	0.07	0	30	15	1013	1.23	7.87	300	152	0.45	0.453333333	0.04		0	0	11.25
--	0.8	0.07	0	30	15	1013	1.23	7.87	300	152	0.46		0.04	0.04	0		11.5
--	1.28	0.09	-0.01	30	15	1013	1.23	10.06	300	152	0.45		0.03		-0.02		15
--	1.29	0.09	-0.01	30	15	1013	1.23	10.06	300	152	0.45	0.45	0.03		-0.02	-0.02	15
--	1.28	0.09	-0.01	30	15	1013	1.23	10.06	300	152	0.45		0.03	0.03	-0.02		15
--	1.75	0.1	-0.01	30	15	1013	1.23	12.05	300	152	0.43		0.02		-0.02		21.5
--	1.75	0.1	-0.01	30	15	1013	1.23	11.98	300	152	0.43	0.43	0.02		-0.02	-0.02	21.5
--	1.74	0.1	-0.01	30	15	1013	1.23	12.05	300	152	0.43		0.02	0.02	-0.02		21.5
--	2.35	0.13	-0.01	30	15	1013	1.23	14.11	300	152	0.42		0.02		-0.01		21
--	2.36	0.12	-0.01	30	15	1013	1.23	14.05	300	152	0.43	0.426666667	0.02		-0.01	-0.01	21.5
--	2.37	0.12	-0.01	30	15	1013	1.23	13.99	300	152	0.43		0.02	0.02	-0.01		21.5
--	3.16	0.15	-0.01	30	15	1013	1.23	15.9	300	152	0.45		0.02		-0.01		22.5
--	3.14	0.15	-0.01	30	15	1013	1.23	15.9	300	152	0.44	0.446666667	0.02		-0.01	-0.01	22
--	3.18	0.15	-0.01	30	15	1013	1.23	15.9	300	152	0.45		0.02	0.02	-0.01		22.5
--	3.87	0.19	-0.01	30	15	1013	1.23	17.79	300	152	0.44		0.02		-0.01		22
--	3.76	0.18	-0.01	30	15	1013	1.23	17.88	300	152	0.42	0.423333333	0.02		-0.01	-0.01	21
--	3.72	0.18	-0.01	30	15	1013	1.23	17.93	300	152	0.41		0.02	0.02	-0.01		20.5
--	4.7	0.25	-0.02	30	15	1013	1.23	19.95	300	152	0.42		0.02		-0.01		21
--	4.92	0.26	-0.01	30	15	1013	1.23	19.95	300	152	0.44	0.433333333	0.02		-0.01	-0.01	22
--	4.99	0.27	-0.01	30	15	1013	1.23	20.04	300	152	0.44		0.02	0.02	-0.01		22
--	5.26	0.2	-0.02	30	15	1013	1.23	22.24	300	152	0.38		0.01		-0.01		38
--	5.43	0.23	-0.02	30	15	1013	1.23	22.13	300	152	0.4	0.396666667	0.02		-0.01	-0.01	20
--	5.56	0.24	-0.02	30	15	1013	1.23	22.09	300	152	0.41		0.02	0.0167	-0.01		20.5
--	6.45	0.34	-0.02	30	15	1013	1.23	24.04	300	152	0.4		0.02		-0.01		20
--	6.48	0.33	-0.02	30	15	1013	1.23	24.07	300	152	0.4	0.403333333	0.02		-0.01	-0.01	20
--	6.6	0.35	-0.02	30	15	1013	1.23	23.97	300	152	0.41		0.02	0.02	-0.01		20.5
--	7.87	0.34	-0.02	30	15	1013	1.23	25.99	300	152	0.42		0.02		-0.01		21
--	7.75	0.31	-0.02	30	15	1013	1.23	26.15	300	152	0.4	0.406666667	0.02		-0.01	-0.01	20
--	7.63	0.3	-0.02	30	15	1013	1.23	26.12	300	152	0.4		0.02	0.02	-0.01		20

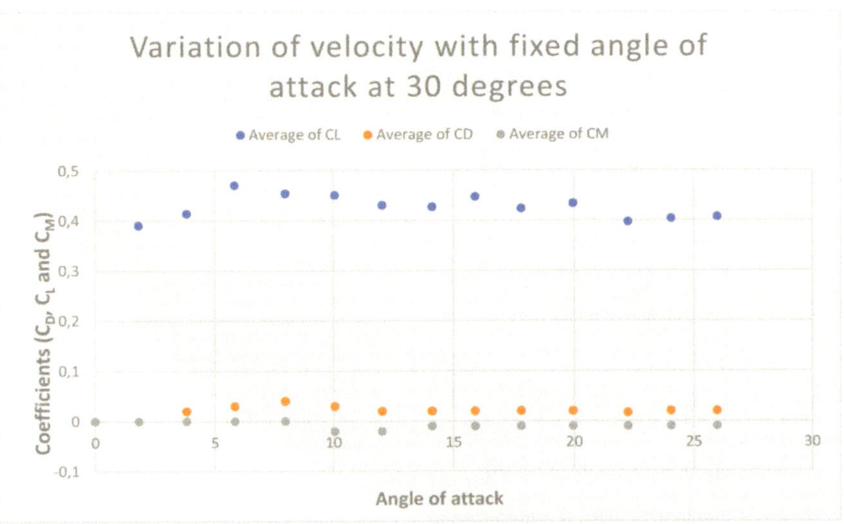

Based on this graph, we can immediately note that an angle of attack of 30 degrees is much better for generating lift, because C_L fluctuates only slightly for varying windspeeds and stays between 0.4 and 0.475 throughout the range of different airspeeds ranging from 0 m/s to 26 m/s.

Coupled with the high C_L values that result in more lift being produced is the very low C_D values throughout the experiment. C_D stays below 0.5 throughout the entire experiment, which means that the drag produced is minimal.

Based on these 2 angles of attack (10 and 30 degrees), it can be concluded from the results that as the angle of attack increases, windspeed becomes a less important factor. In other words, increasing the angle of attack eliminates windspeed as a factor for lift and drag forces.

The C_M values are very close to zero for the duration of the experiment, which means that just like in the previous graph, the pitch attitude was stable.

Calculating the values theoretically

I shall use 3 equations to calculate C_D, C_L and C_M. The equations are as follows:

C_L equation: $c_L = \dfrac{2L}{\rho S v^2}$ C_D equation: $c_D = \dfrac{2D}{\rho S v^2}$ C_M equation: $C_M = \dfrac{M}{qSc}$

Where L = lift, S = aerofoil area, M = pitching moment, q = dynamic pressure and c = aerofoil chord length

S = 300 x 152 = 45 600 mm² = 0.0456 m²

Example of theoretical calculations from experiment 1

AoA 12 deg. $- c_L = \dfrac{2(2.43)}{1.225(0.0456)(20)^2} = 0.22$

$$c_D = \dfrac{2(7.37)}{1.225(0.0456)(20)^2} = 0.66$$

AoA 30 deg. $- c_L = \dfrac{2(10.74)}{1.225(0.0456)(20)^2} = 0.96$

$$c_D = \dfrac{2(7.57)}{1.225(0.0456)(20)^2} = 0.68$$

The calculated values for 12 degrees AoA are 0.22 and 0.66 for C_L and C_D respectively, whereas the experimental values are 0.23 and 0.7.

The calculated values for 30 degrees AoA are 0.96 and 0.68 for C_L and C_D respectively, whereas the experimental values are 1.07 and 0.76.

This shows us that there's a slight error between the calculated values and experimental values, and this error is smaller for smaller AoA. One of the reasons for these disparities is that in calculations, it's assumed that the conditions are perfect, or ideal, when in reality conditions are never perfect. For example, humidity is not accounted for in theoretical calculations, whereas in actual situations humidity is a factor that affects the results.

Additionally, all other weather patterns such as rain, hail, fog, etc. can't be replicated in a wind tunnel, so experimental values will always differ from the theoretical values.

Task 2

Wing planform geometry and its effects on lift and drag generation

A planform is a wing's shape when viewed top-down. The wing planform deals with airflow in 3 dimensions. The main factors affecting a wing planform design are taper ratio, aspect ratio and sweepback.

Aspect ratio is "the ratio of wingspan to wing chord". Taper ratio is "the ratio of tip chord to root chord", and sweepback is "the rearward slant of a wing, horizontal tail, or other airfoil surface". (flightliteracy.com, 2015)

Increasing the aspect ratio while maintaining constant velocity reduces the amount of drag produced, and this effect increases the larger the angle of attack is. Thus, a higher aspect ratio improves an aircraft's climbing performance. However, for a higher aspect ratio to be achieved, the aircraft's wingspan must also increase, thus adding extra weight to the aircraft for carrying the same load.

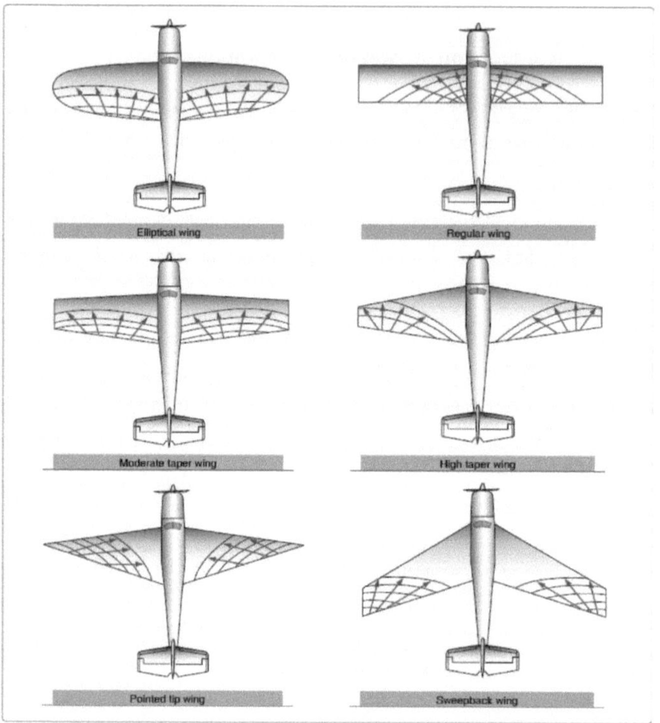

Figure 2 Different types of wing planform in small aircraft (flightliteracy.com, 2015)

On the other hand, decreasing the aspect ratio corresponds to an increase in drag, so a compromise between the two must be reached, and depending on the situation, some designers prioritize one thing over the other.

Tapering is another way of decreasing drag forces. Tapering refers to reducing the chord length from the wing's root to its tip. As well as a decrease in drag, tapering increases the lift produced, and this is more notable at higher velocities. Additionally, a reduced chord length reduces the wing weight. When the taper ratio is combined with the aspect ratio in terms of design, a better compromise can be reached because now the wing can be longer without adding more weight.

A very low aspect ratio results in high wing loading and stall speeds. When the wings are swept back in combination with a low aspect ratio, the flying qualities differ greatly compared to conventional high aspect ratio aircraft. To maneuver such aircraft effectively, very precise flying techniques are required, even more so at lower velocities, whereas aircraft with higher aspect ratios mitigate some of the improper techniques and mistakes, thus making flight in high aspect ratio aircraft more stable. (flightliteracy.com, 2015)

Boundary layer effects on lift and drag

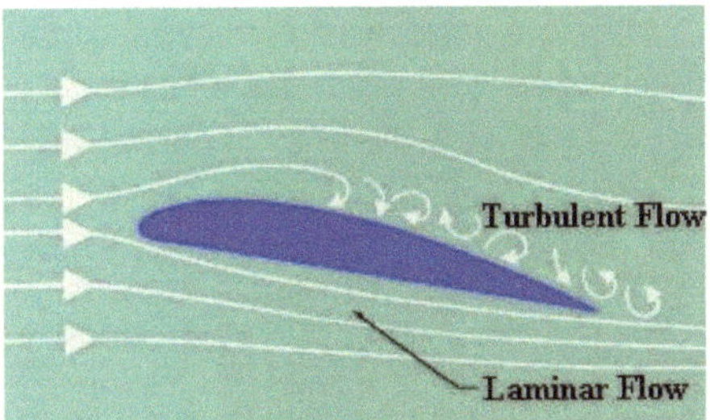

Figure 3 Laminar and turbulent flow in the boundary layer (pilotfriend.com, n.d.)

The boundary layer is very thin layer over a wing's surface, and it consists of a layer of motionless particles on the aerofoil's surface. These particles slow down other molecules just above them, which in turn slow down the molecules above them and so on, until the last layer, where the molecules travel at the airflow speed. The boundary layer can severely affect the aerodynamic forces of lift and drag by increasing drag and decreasing lift if it's not accounted for. The airflow over the aerofoil is turbulent if there aren't any extra design features added. Turbulent flow can be completely eradicated by installing vortex generators on the aerofoil surface.

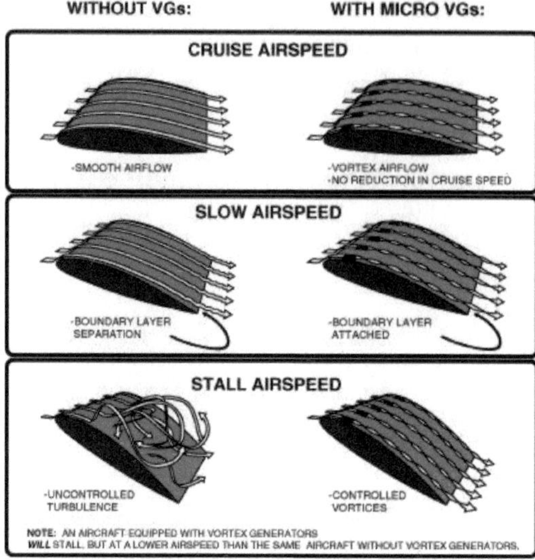

Figure 4 Effect of adding vortex generators (pilotfriend.com, n.d.)

As the boundary layer shifts towards the middle of the wing it begins losing speed due to skin friction drag. The boundary layer at its transition point transitions from laminar, where the velocity varies uniformly as you move away from the surface of the body, to turbulent, where the velocity is defined by erratic swirling flows within the boundary.

The airflow outside the boundary layer reacts to the shape of the boundary layer's edge in the same manner that it does to the aerofoil's physical shape. (pilotfriend.com, n.d.)

Based on the conservation of mass in 3 dimensions, variations in velocity in the streamwise direction results in the velocity changing in other directions as well. A small component of each velocity is perpendicular to the surface, and these displace the flow above them. Boundary layer thickness can be defined as the amount of this displacement, i.e. the airflow velocity, combined with the distance it traveled across the aerofoil, determine whether the airflow is laminar or turbulent. The displacement thickness is dependent on Reynolds number, which is "the ratio of inertial forces to viscous forces". The equation is given by:

$$R_e = \frac{v\rho l}{\mu}$$

For lower Reynolds numbers, the boundary layer airflow is laminar, and the streamline velocity varies uniformly as one moves away from the aerofoil. (grc.nasa.gov, 2015)

Atmospheric events

Severe air turbulence

Turbulence occurs when different air currents mix and disrupt the otherwise smooth airflow. For example, when cold air currents mix with warmer air currents, or airflow is disturbed by cliffs and mountains at lower flight altitudes, or during thunderstorms, then the air will be "bumpy", and the aircraft will experience rolling motions due to tilting of the wings.

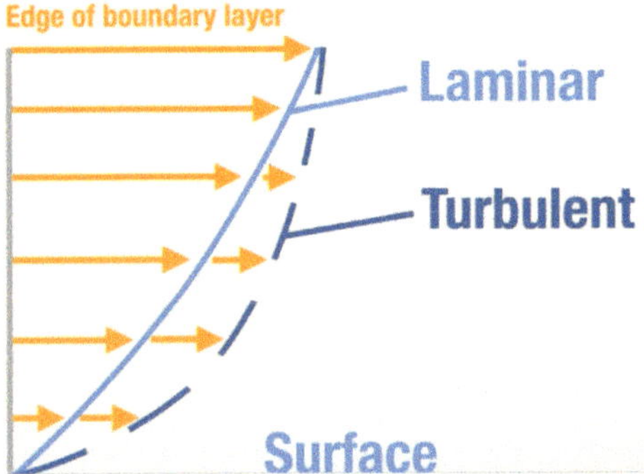

Figure 5 Velocity profiles for laminar and turbulent airflow (Udris, 2016) (boldmethod.com)

Turbulence results in a thicker boundary layer, thus generating more skin friction drag, which further affects airflow speed in the lower region of the turbulent boundary layer by reducing the airspeed just above the surface, unlike in laminar flow.

Pressure drag is more significant on larger aircraft than skin friction drag, mainly on the fuselage and nacelles. Overall drag is reduced by considering the fact that a turbulent boundary layer has more energy to oppose adverse pressure gradients. The overall drag reduction is thus a result of purposely forcing the boundary layer to turn turbulent over the fuselage, but not the wings. (Udris, 2016) (boldmethod.com)

Frost and ice accretion

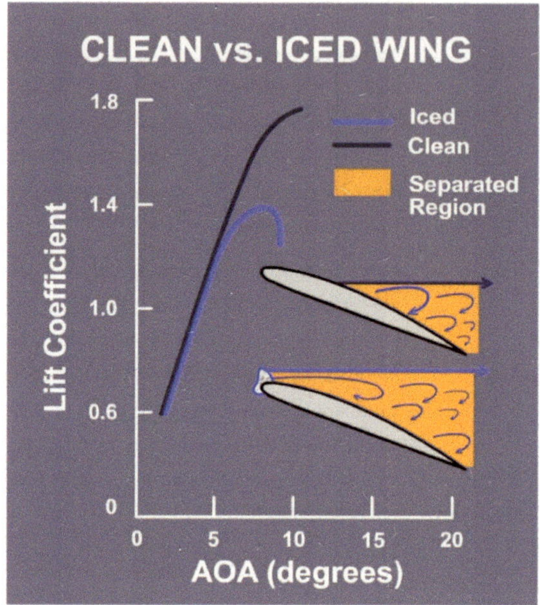

Figure 6 Airflow around clean vs iced aerofoils (Wadel, 2016) (aircrafticing.grc.nasa.gov)

Ice build-up over aerofoils, ailerons and elevators can result in changes to the aerodynamic forces of lift and drag, thus resulting in roll and pitch upsets due to stalling of the tail and wings.

As seen in the picture on the left, the angle of attack affects the stalling of the wing during ice build-up. (Wadel, 2016) (aircrafticing.grc.nasa.gov)

This reduces lift by a steep amount (blue curve in the graph). Furthermore, flow separation can occur prior to artificial stall warning, thus causing the aircraft to stall without warnings or indications appearing to the pilot.

Aero-elasticity effects

Wing torsional divergence and flutter

Aircraft aren't completely rigid objects; the wings are fairly flexible compared to the fuselage, which is relatively rigid. An example is wing flexing relative to the fuselage. This is because the wings bend more at the tips compared to the roots. This is called divergence, whereby the moment produced by the air load is greater than the structural torsional stiffness of the wing, which would result in the wing being twisted off the fuselage.

The threshold speed for such a failure is known as the divergence speed, which should be much higher than the aircraft operating speeds. Swept forward wings are the most susceptible to this scenario because they have low divergence speeds.

Another scenario is when the wing rotates, or twists, around its stiffness axis, or spar. This is called flutter. In this case there is a synchronised interaction between the two modes, so that in one mode the energy is absorbed from the airflow to increase the other mode's amplitude. Each mode's frequency then converges to the same value at this point, so that only one combined mode is possible. The wing can draw energy from the airflow and act as an ever-increasing flexure in bending and torsion until enough displacement is reached to break the wing.

When the airflow is increased to the critical point which results in this failure, it's known as flutter speed. Just like with divergence speed, the flutter speed should be much higher than the aircraft's operating range. (aerodynamics4students.com, 2019)

Controls reversal

This is an unfavourable effect on the aircraft's controllability, whereby the aircraft's flight controls are reversed in such a way that's not natural for the aircraft operation. Pilots might be unaware of this, so rather than having to roll to the left, for example, they would have to push the control stick in the opposite direction. The primary cause of this scenario is high speed flight, i.e. supersonic flight.

Another cause can be wing twisting due to the ailerons generating sufficient force to twist the wing, as explained in the aeroelasticity section. This can happen in supersonic flight because of the high compressibility effects generated.

Calculations for level flight

Q1. (i)

a) $W = 280(1000)(9.81) = 2\,746\,800$ N

b) $\dfrac{L}{D} = \dfrac{C_L}{C_D} \Rightarrow L = W \Rightarrow c_L = \dfrac{2(2\,746\,800)}{1.225(520)(111.11)^2} = 0.699$

$$C_D = C_{D_o} + \frac{(C_L)^2}{\pi ARe} \Rightarrow C_{D_o} = 0.036,\ AR = \frac{L^2}{S}\ C_D = 0.036 + \frac{0.699^2}{\pi\left(\frac{60^2}{520}\right)0.7} = 0.068$$

$$\frac{L}{D} = \frac{0.699}{0.068} = 10.279 = 10.28$$

c) Thrust required $\Rightarrow \dfrac{W}{L/D} = \dfrac{2\,746\,800}{10.279} = 267\,224$ N

d) Thrust available $\Rightarrow T = 4(205\,000 - 681.53(111.11) + 2.236(111.11)^2) = 627\,520$ N

(ii)

Stall speed (km/h)	Thrust required	Thrust available
200	539518.9338	696153.8272
300	303682.7977	654934.4444
400	267588.0668	627517.5309
500	301841.3627	613903.0864
600	374514.4494	614091.1111
700	474667.3619	628081.6049
800	597730.35	655874.5679
900	741520.3109	697470
1000	904887.424	752867.9012

Aircraft weight (N)	Cl	Cd	L/D	Pi
2746800	2.79	0.54883558727524	5.091202	3.141593
	1.24	0.13730085674573	9.044964	
	0.70	0.06805222420470	10.26503	e
	0.45	0.04912859103425	9.100144	0.7
	0.31	0.04233130354661	7.334296	
	0.23	0.03941747996518	5.786789	Ar
	0.17	0.03800326401279	4.595383	6.923077
	0.14	0.03725062786106	3.704282	
	0.11	0.03682053693964	3.035516	

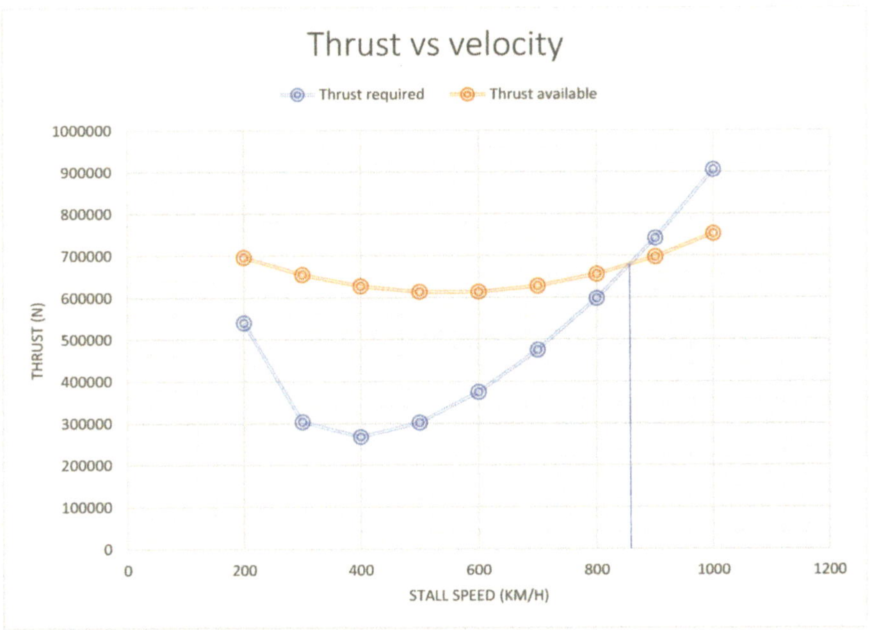

Estimated maximum velocity = 860 m/s.

Calculations for maneuvers

Q2. (i)

a) m = 7000 kg s = 50 m² v = 425 knots ⇒ 218.639 ms⁻¹ bank angle = 30°

$$n = \frac{1}{\cos(30)} = \cos(30)^{-1}$$

$$r = \frac{v^2}{g\sqrt{n^2-1}} \Rightarrow r = \frac{218.639^2}{9.81\sqrt{(\cos(30))^{-1})^2-1}} = 8440 \text{ m}$$

$$a = \frac{v^2}{r} \Rightarrow a = \frac{218.639^2}{8440} = 5.66 \text{ ms}^{-2}$$

$$F = ma \Rightarrow F = 7000(5.66) = 39\,620 \text{ N}$$

b) Wing loading $\Rightarrow F = \frac{mv^2}{r} = \frac{7000(218.639)^2}{8440} = 39\ 647$ WL $= \frac{39647}{50} = 792.9$ Nm^{-2}

(ii)

   a.   r = 2000 m $a = \frac{v^2}{r} \Rightarrow a = \frac{218.639^2}{2000} = 23.9$ ms^{-2}

   b.   (i) $F = \frac{mv^2}{r} = \frac{7000(218.639)^2}{2000} = 167\ 310$ N WL $= \frac{167\ 310}{50} = 3346$ Nm^{-2}

          (ii) $F = 167\ 310 - 68\ 670 = 98\ 640$ N WL $= \frac{98640}{50} = 1973$ Nm^{-2}

Forces on an airplane

Lift: Angle of attack is increased, along with additional thrust from the engines, to overcome the weight.

Drag: This pertains to various resistive forces against the aircraft's motion.

Thrust: Engines use a propulsion system to overcome drag by generating a force called thrust. The thrust force direction depends on how the aircraft engines are placed.

Aircraft maneuver envelope

This refers to an aircraft's flight envelope, or in other words the design capabilities in terms of airspeed and load factor or altitude. Before they can be certified as airworthy, aircraft should meet those performance and design specifications.

A typical maneuver envelope is a speed-versus-load factor graph. Speeds are dictated by the efficiency of the aircraft in handling and the required cruise activity. Charging considerations are set as the vehicle's restricting design criteria. The aircraft and all of its components must be configured to function safely inside this envelope at any level. (aerodynamics4students.com, 2015)

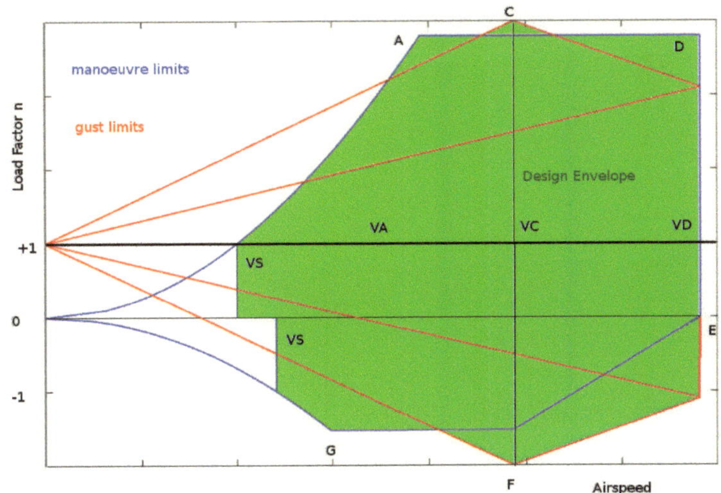

Figure 7 VN graph (aerodynamics4students.com, 2015)

Limit load

The VN diagram is also known as a velocity vs load factor chart, which shows how much load factor can be safely achieved at varying airspeeds. Air is less dense at higher temperatures; therefore, aircraft must fly higher at greater speeds to produce the same amount of lift as in lower altitudes.

The load factor (n) = lift / weight shows the relationship between lift and weight; it's equal to 1 when the aircraft is static on the ground with the only force acting on it being gravity. The load factor can also be defined as a multiple of gravitational acceleration, g. The load factor also hints at the types of forces that the aircraft structure can withstand in flight.

The VN diagram shows the structural load limits as a function of airspeed. The maneuver envelope is defined during the preliminary design phase. This graph shows the limits of an aircraft's performance. The definition as well as analysis of a VN graph is crucial during aircraft design, since it affects the aircraft's operation. Structural damage can occur in case winds buffering the aircraft force it out of the safe region of the envelope. (uavnavigation.com, 2020)

$v_p = v_s\sqrt{n_{limit}}$ where v_p = maneuver speed, v_s = stall speed and n_{limit} = limit load factor

36

This number is valid only for the computed gross weight, altitude and configuration. (aviationchief.com, 2018)

Bibliography

aerodynamics4students.com, 2015. *FLIGHT ENVELOPE.* [Online]
Available at: http://www.aerodynamics4students.com/aircraft-performance/flight-envelope.php
[Accessed May 2020].

aerodynamics4students.com, 2019. *AEROELASTICITY - Wing – Flutter and Divergence..*
[Online]
Available at: http://www.aerodynamics4students.com/aeroelasticity/
[Accessed May 2020].

aviationchief.com, 2018. *Operating Flight Strength (V-g / V-n Diagrams).* [Online]
Available at: http://www.aviationchief.com/operating-flight-strength-v-g--v-n-diagrams.html
[Accessed May 2020].

flightliteracy.com, 2015. *Effect of Wing Planform.* [Online]
Available at: https://www.flightliteracy.com/effect-of-wing-planform/
[Accessed 2020].

grc.nasa.gov, 2015. *Boundary Layer.* [Online]
Available at: https://www.grc.nasa.gov/WWW/K-12/airplane/boundlay.html
[Accessed 2020].

mdp.eng.cam.ac.uk, n.d. *PROPERTIES OF THE ATMOSPHERE.* [Online]
Available at: http://www-
mdp.eng.cam.ac.uk/web/library/enginfo/aerothermal_dvd_only/aero/atmos/atmos.html
[Accessed April 2020].

nasa.gov, 2017. *Earth's Atmospheric Layers.* [Online]
Available at: https://www.nasa.gov/mission_pages/sunearth/science/atmosphere-layers2.html
[Accessed 2020].

pilotfriend.com, n.d. *Wing Boundary Layer.* [Online]
Available at: http://www.pilotfriend.com/training/flight_training/aero/boundary.htm
[Accessed 2020].

Sharp, T., 2017. *Earth's Atmosphere: Composition, Climate & Weather.* [Online]
Available at: https://www.space.com/17683-earth-atmosphere.html
[Accessed 2020].

SkyBrary, 2017. *Drag.* [Online]
Available at: https://www.skybrary.aero/index.php/Drag
[Accessed May 2020].

The Engineering ToolBox, n.d. *U.S. Standard Atmosphere - Properties of US standard atmosphere ranging -5000 to 250000 ft altitude.* [Online]
Available at: https://www.engineeringtoolbox.com/standard-atmosphere-d_604.html
[Accessed May 2020].

uavnavigation.com, 2020. *Flight Envelope.* [Online]
Available at: https://www.uavnavigation.com/support/kb/general/general-system-info/flight-envelope
[Accessed May 2020].

Udris, A., 2016. *How Turbulent Is The Airflow Over Your Wings?.* [Online]
Available at: https://www.boldmethod.com/learn-to-fly/aerodynamics/boundary-layer/
[Accessed 2020].

Wadel, M., 2016. *Handling Effects Aerodynamics of Icing.* [Online]
Available at: https://aircrafticing.grc.nasa.gov/1_1_3_3.html
[Accessed 2020].